Zoey's Space Adventures
Copyright © 2022 by Kesha Robinson. All rights reserved.

No part of this book may be reproduced in any form without written permission from the copyright owner.

Written by Zoey & Kesha Robinson
Illustrated by Nadia Rajput

Zoey's Space Adventures

By Zoey & Kesha Robinson

Dedicated to my grandma,
Doris Bishop at the FAA Commercial
Space Transportation Team.

JUPITER

MERCURY

Next up is Mercury! Did you know that Mercury is the smallest planet in the solar system? It is the closest planet to the sun and also has the thinnest atmosphere.

URANUS

If you guessed Venus, you're right! Did you know that Venus is the second planet from the sun? It is the closest planet to Earth and known as Earth's twin. It is also the hottest planet in the solar system and the first planet to be explored by a spacecraft.

VENUS

Now it's time to explore Neptune. Did you know that Neptune is the farthest known planet in the solar system? Neptune actually has six rings but they are very hard to see. It is also very windy. Wind speeds are among the fastest recorded in the solar system.

NEPTUNE

Wheeee!!!
Now let's take a moment to play and talk about gravity. Gravity is an invisible force that pulls objects toward each other. Astronauts feel weightless when there is nothing opposing the force of gravity.

Now let's identify asteroids, meteoroids, and comets! An <u>asteroid</u> is a rocky object that orbits the sun. A <u>meteoroid</u> is a small space rock moving through the solar system. These are smaller than asteroids. A <u>comet</u> is a small object made of ice and dust that orbits the sun. It has a long tail of gas.

METEOROID

ASTEROID

COMET

EARTH

Yayyy!! It's planet Earth. This is where we live! Did you know that Earth is the third closest planet to the sun? It is 71% covered by water and 29% covered by land. It is the fifth largest planet in the solar system.

Last is Saturn. Did you know that Saturn is the second largest planet in the solar system? It has at least 62 moons and is most famous for its beautiful rings. This is another windy planet and it has the lowest density out of all of the planets in the solar system.

SATURN

> It is so cool to see all of the planets together. Each of them are so unique. Look at all of the different sizes and colors.

Fun Fact

Did you know that Pluto was once considered the ninth planet in our solar system? It has now been re-classified by the IAU as a dwarf planet. A dwarf planet is a small body similar to a planet, but lacking some of the criteria. Pluto is one of several dwarf planets.

Pluto

Learning and exploring can be so much fun! Be sure to embrace each new adventure and enjoy!!

About the Author

Zoey Robinson is a seven-year old best-selling author, influencer, kid entrepreneur, and Certified Game & App Developer. Zoey currently hosts her own YouTube & Television show, Zoey TV and has released 10 books since January 2020. Zoey has been featured nationally & internationally in magazines, radio shows, podcasts, Roku TV, Spotify, local news, and other media platforms. Some of her recent achievements include recognition by United States Senators Chris Van Hollen and Lena C Taylor, Entrepreneur of the Year Award, Awesome Young Authors' Award, Top 10% in Public Speaking & Creative Writing, and #1 on the global leaderboard in coding, along with many other honors and awards. As part of her literacy campaign, she does appearances, including school visits and participates in read-a-thons to promote positivity & literacy. Zoey is on a mission to spread positivity, creativity and joy throughout the world!!

👍 **Book Signings**
👍 **Book Readings**
👍 **Interviews**
👍 **School Visits**
👍 **Vending Events**
👍 **Appearances**

Books, MERCH
AND MORE

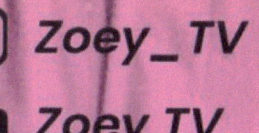
Zoey_TV
Zoey TV

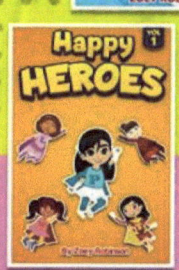
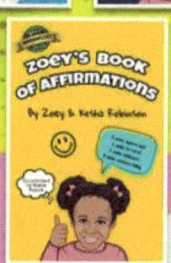
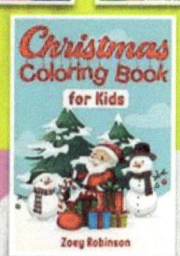
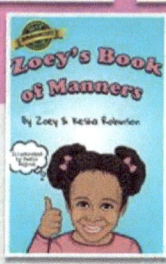

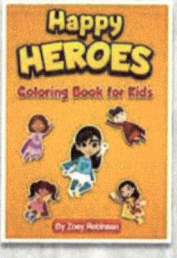

www.zoeytv.com

www.ingramcontent.com/pod-product-compliance
Lightning Source LLC
Chambersburg PA
CBHW041943240526
45473CB00033B/466